STEAM ROLLERS IN PRESERVATION

MALCOLM BATTEN

AMBERLEY

First published 2024

Amberley Publishing
The Hill, Stroud
Gloucestershire, GL5 4EP

www.amberley-books.com

ISBN 978 1 3981 2105 8 (print)
ISBN 978 1 3981 2106 5 (ebook)

British Library Cataloguing in Publication Data.
A catalogue record for this book is available from the British Library.

Typesetting by SJmagic DESIGN SERVICES, India.
Printed in the UK.

Contents

Introduction

The term 'steam roller' has become part of the English language, rather like Biro or Hoover. The first recorded British use of a steam engine for road rolling is in 1859 when William Bray of Folkestone used a traction engine to tow a roller. In 1865 Thomas Aveling first fitted a traction engine with smooth wheels for rolling the road surface. These rear wheels were 7 feet in diameter and 3 feet wide. The machine weighed around 20 tons and was tried out in Hyde Park, London. Other machines followed including a machine for Liverpool Corporation in 1867, which had chain-driven rolls mounted either side of the boiler and a massive full-width rear roll below a 500-gallon water tank, which was steered by a ship's wheel on the driver's platform. This engine weighed around 30 tons in working order, which proved too heavy, crushing rather than rolling the stone. However, development progress was rapid and by the end of the 1870s Thomas Aveling perfected the most common three-point design with a wide two-piece roll mounted in a steering fork at the front and a roll either side at the rear. His company, Aveling & Porter, became the largest manufacturer of steam rollers, making about two-thirds of the approximately 6,000 produced for the home market, although most other traction engine builders also produced designs – some quite innovative.

The customers for steam rollers fell into two main groups – local authorities and contractors. The local authorities could range from urban and rural district councils to county councils and county borough councils. However, in 1930 county councils became responsible for roads so the rollers from the urban and rural district councils were absorbed by their county councils.

The rollers came in various sizes and weights, usually between 6 and 10 tons, but could be up to 16 tons unladen. Most would have a large solid flywheel, intended to give smooth running. There were both single cylinder and compound engines, but the single-cylinder layout remained most popular, and some of the last built rollers were of this type. Steering was usually by chains to the front roll. As well as rolling the road, many were fitted with a scarifier behind the rear rolls. This was a large spike that could be wound down into a pickaxed hole in the road to rip up the old surface before the new surface was laid. Customers could specify the type of scarifier fitted and makers included Price, Morrison and Allen. There would be scrapers fitted to remove any road metal caught on the rolls. Some rollers also had sprinklers to wet the road surface and reduce dust.

In earlier times the road surface was made of crushed stones compacted into place by the roller. When asphalt and tar-based materials came into use as a road surfacing this caused a problem. Traction engines and conventional rollers need to stop to change gear and reverse and this led to a degree of 'dwell' or sinking into the surface while this took place, leading to an uneven surface. Tandem rollers with equal weight on each roll, vertical boilers, an instant-reverse valve gear and no flywheel were developed by Aveling & Porter and Marshall, amongst others, to get round this problem. Later, Wallis & Steevens introduced their 'Advance' design, which featured two high-pressure cylinders, one for forward and one for reverse with an instant reverse capability. Some rollers were also equipped for tar spraying. Alternatively, many contractors adapted steam wagons for tar spraying, their boilers keeping the tar hot for spraying.

Many manufacturers' steam rollers shared component parts with their other designs and engines would sometimes be converted to or from rollers during their working life. However, there were also some built that were designed to be convertible and supplied with both road wheels and rolls with their front fork bracket, which could be fitted to a bolted flanged ring in front of the smokebox. Such engines were aimed at large estates where there might be a need for a versatile engine to haul produce to market, drive farm machinery on the belt from the flywheel and also undertake maintenance of the estate roads. The changeover could be effected in around a day with jacks and lifting gear.

Internal combustion motor rollers first appeared around 1900 in France. The first UK application came in 1905 when J. G. Allen & Co. of the Irish Road Engineering Works produced a 15-ton roller with a single-cylinder petrol engine. Motor rollers were developed especially by Barford & Perkins of Peterborough and Thomas Green of Leeds particularly for the smaller roller market of sports clubs and country estates. Aveling & Porter produced their first motor roller in 1913 and their rollers were later produced in parallel with steam rollers until eventually succeeding them. Barford & Perkins were the first to fit a diesel engine in 1927. The last steam rollers for the home market were built by Aveling-Barford in 1948.

Steam rollers were the last type of steam-powered road vehicle in general use, some remaining in service until the mid-1960s. Consequently, many hundreds have survived into preservation and can regularly be seen at rallies and other events. The largest rally is the Great Dorset Steam Fair and at the 2013 show there was a special roller theme with over 140 steam, petrol and diesel rollers attending.

While some rollers passed directly into preservation, others were acquired from scrap merchants. Some local authorities installed rollers (minus their brass fittings, etc) into playgrounds as static climbing frames, having been donated these by Eddison Plant Ltd. When health and safety legislation made such practice unacceptable, some of these engines were sold into preservation. In more recent years, the number of rollers, and all other types of traction engines, has increased substantially through the repatriation of exported engines that had survived in such places as Australia and South America.

The umbrella organisation for the preservation movement is the National Traction Engine Trust (NTET), originally the National Traction Engine Club. This was set up in 1954 to ensure the preservation of road going steam engines. Their membership comprises individual members and affiliated clubs from throughout the world. They produce a Code of Practice under which approved rallies are held. They set safety standards and provide technical advice, and campaign against any legislation that might adversely affect the preservation hobby. They also secure insurance and other related services for members and rally organisers, support the maintenance of records of engine manufacturers, and provide facilities to teach basic skills in engine management including attracting junior members through the Steam Apprentice Club, founded in 1979.

The Road Roller Association was formed in 1974 and this aims to encourage the preservation of steam rollers, motor rollers and all sorts of road-making equipment. They have created a register of motor rollers and publish a quarterly magazine, *Rolling*.

The *Traction Engine Register*, first published in 1966, is the authoritative listing of all known traction engines in the United Kingdom and Eire, including steam wagons, fire engines and portable engines. There are over 4,000 engines listed in total, and this is updated every four years. The number has grown over the years due to exported British-built engines being repatriated from around the world, some imports of foreign made engines, and even the construction of 'new-build' engines, although this practice has not stretched to rollers (yet).

This book should perhaps be subtitled 'and former steam rollers'. Some engines that were built as rollers were converted to tractors during their working lives, or vice versa, and some have since been changed back. Unfortunately, many more rollers have been converted into tractors or showman's tractors since preservation, including at least thirty Fowler rollers. This is a practice that the National Traction Engine Trust does not condone while recognising the right of owners to do as they please with their own property. They issued a statement: 'One of the major objectives of the NTET is to preserve our steam heritage, including the original design of our road steam vehicles, their use and maintenance. Therefore, the Trust does not support in any way the conversion of any steam road vehicle into a form which that particular machine never had during its working life.'

This book illustrates many of the examples from different makers that can be seen in Britain and tells of their history into preservation.

All photographs are by the author except where stated.

1
Flashback – Working Days

Some steam rollers remained in commercial use into the 1960s but that did not guarantee their preservation. Aveling & Porter No. 4855 (ME 3266) was No. 5 in the fleet of Rowley Plant Hire Co. Ltd when seen at their depot at Beddington Lane in December 1950 but did not survive. This was registered No. 64 with Middlesex County Council. *Photo by John Meredith/Online Transport Archive*

Also with Rowley Plant Hire Co. Ltd and seen under repair on the same day is Marshall roller No. 78724 (OPC 765). This was preserved but has since been converted to a tractor in 1975 – a fate that has befallen many rollers. *Photo by John Meredith/Online Transport Archive*

Left: Wallis & Steevens 'Advance' roller No. 8005 (OU 2684) at work alongside the tram lines in Westminster Bridge Road, London, on 9 November 1950. This was with H. Woodham & Sons Ltd. This has also survived into preservation. *Photo by John Meredith/Online Transport Archive*

Below: A local council-owned roller, Aveling & Porter No. 14008 (VG 2269) of 1929, was bought by the City of Norwich as their No. 7 and is seen in Carrow Road on 28 June 1956. This roller is now preserved in Norfolk as part of the Thursford Collection. *Photo by John Meredith/Online Transport Archive*

One of the many rollers operated by the contractors Eddison Steam Roller Co. of Dorchester, Aveling & Porter No. 10422 (FX 9699) taken in September 1949. The Aveling & Porter rampant horse emblem has gone from the headstock. This engine did not survive. *Photo by John Meredith/Online Transport Archive*

Looking past its best and probably disused at the time when photographed in June 1956 is Wallis & Steevens No. 7247 (HO 6417) of 1912. Latterly with F. R. Godden at Southwater, this did pass into preservation and was listed as being based in Cornwall in 2020. *Photo by John Meredith/Online Transport Archive*

2
Steam Rollers by Make

Allchin (William Allchin Ltd, Northampton)

Two steam rollers from the company of William Allchin survive. Both worked for the County Borough of Northampton where the Allchin works was located. No. 1311 (NH 3416) was built in 1900 as fleet No. 1 and cost £400 new. Both survived in council ownership after withdrawal. In 2009 the Borough Council offered the unrestored NH 3416 to the Northampton & Lamport Railway on condition that the roller would be restored and maintained at the preservationist's expense and would be painted in Council livery and made available for certain civic events. The 10-ton Allchin was restored in 2012 and is seen in 2014 at the Bedfordshire Steam & Country Fayre at Old Warden now named *Northampton*. Note the massive size of the flywheel.

No. 1187 of 1901 was No. 2 in the Northampton fleet. First purchased for preservation in 1961, it was acquired by the present owner in 2017 and since restored to the original specification, returning to the scene in 2022. As such, this rear view shows it at the 2023 Bedfordshire Steam & Country Fayre held at Turvey House.

Armstrong-Whitworth (Sir W. G. Armstrong-Whitworth Ltd, Manchester)

The firm of Sir W. G. Armstrong-Whitworth Ltd, Manchester, made a small number of steam rollers in the 1920s, of which seven survive in the United Kingdom. All but one of these are 10-ton rollers, the other being 12 tons, and all are compounds with piston valves. Oldest of the survivors is BD 7511, first registered in 1923 for Kettering Urban District Council. Note the extra water tank, mounted under the boiler on this engine. Taken at the Gloucestershire Steam & Vintage Extravaganza in 2007.

There was no canopy fitted on No. 10R19 (FJ 2795) when seen at Redhill in 1990. All right on a hot sunny day like this, but not so nice for the crew when it rained.

DX 4602 of 1924 is preserved at the East Anglia Transport Museum at Carlton Colville near Lowestoft. This was new to Ipswich Docks Commission and worked for some forty years. It was keeping company with another local roller, Aveling & Porter No. 11648 of 1926, which worked for the local Borough of Lowestoft.

NE 2271 *Badger* of 1925. This resides in Sussex and was taken at Ardingly in 1984.

Aveling & Porter (Aveling & Porter Ltd, Rochester, Kent)

Aveling & Porter Ltd, Rochester, Kent, became the largest manufacturer of steam rollers and some 600 survive in preservation. In the 1920s Aveling & Porter joined with other companies including Burrell, Garrett and Barford & Perkins to form the Agricultural & General Engineers (AGE). This later collapsed and when Thomas Lake Aveling, the founder's son, died in 1931 the company was rescued by Edward Barford and became Aveling-Barford Ltd in 1932, but engines continued to be badged as Aveling & Porter until 1937.

One of the oldest Aveling & Porter rollers being rallied is No. 2754 (KE 1999) of 1890. This is an R15 model, weighing 15 tons and with a single cylinder. This was seen at Horstead Keynes, home of the Bluebell Railway on 31 July 1994.

Highest numbered Aveling & Porter engine in the *Traction Engine Register* is No. 14187 of 1937 named *Ultimus*. The AD design was built from 1930 onwards and this 10-ton model was built by Aveling-Barford and supplied new to Haslemere UDC. It later saw service with Arnold & Nathan Ltd, Paddock Wood. This was recorded at the Knowl Hill Rally in August 1993.

Aveling & Porter built rollers in a variety of different weights. At the Bedfordshire Steam Show in 2003 these three examples were posed together to aptly demonstrate this. On the right is No. 7385. FC 4016 was a 4-ton R4 model made especially narrow for the requirements of the City of Oxford to work in some of their narrow lanes. In the middle is No. 10647 (PD 6727), a class E 10-ton roller. On the left is No. 8717 (SX 923) *Jupiter,* a 15-ton BHP class roller which worked in Scotland until 1950.

This rear view of No. 7385 at St Albans in 2014 emphasises the small size and narrow width of the Oxford roller. New in 1911, this was replaced by a diesel roller in 1938 and sold to J. Allen of Cowley. It then moved to Taylor Bros of Essex and used for trench replacement and on airfields during the Second World War.

Often at rallies, rollers will be displayed in a road making display setting. At Barleylands in 2006, No. 11064 *Cestria* of 1924 is belted up to a stone crusher. Note the scarifier. Some vintage barriers and paraffin lamps complete the scene. This roller was new to an owner at Chester-le-Street, Durham.

The roller crew would have travelled from job to job with a living van and probably a water cart to carry extra water in case there was not a convenient hydrant to top up supplies. C class 8-ton roller No. 11347 of 1925 has a living van and water cart in tow in this scene from 1979. The location is Hatfield House, Hertfordshire.

While rollers may be displayed in a road-making display setting at rallies, this view at Chatham Dockyard probably recreates a typical council yard scene better than could be achieved on any field site. No. 14073 (FV 1326) was new to Blackpool Corporation in 1930 and is a 10-ton AC model.

While a working scene might be recreated on a rally field, in their working life rollers would be more associated with roads than fields. At the former Birmingham Museum of Science & Industry in Newhall Street, a road run was held around the back streets. No. 11520 (MK 5128) a 10-ton E class roller from Wolverhampton, heads through a scene probably little changed since its working days. May 1985.

Engines line up opposite the Science Museum ready for the road run. In 1984 we see No. 14062 of 1930, a 12.5-ton compound cylinder roller of class AF. This is a roller which I may well have seen late in its working life as it was bought by the County Borough of East Ham, where I grew up.

Meanwhile, inside the Science Museum resident No. 2992 (AB 9331) was being steamed. Note the smoke extractor extension from the engine's chimney. I wonder what the health and safety people would say about such practices these days! The museum closed in 1997 and has since been replaced by a new interactive museum complex, known as 'Thinktank', which opened on 29 September 2001 at Millennium Point, Curzon Street, after five years of construction. Unfortunately, the steam line-up and road run have not been continued at the new location, bringing an end to what was probably the only regular opportunity to photograph traction engines on a run in an industrial street setting. This is fitted with a Morrison three tine patent scarifier.

While most rollers entered at steam rallies are fully restored, sometimes an unrestored engine may be on show, illustrating the amount of work that will need to be undertaken to restore this to full working order. Such was the case with No. 8290, a BS class 10-ton roller of 1904. This was at Harting, West Sussex, in July 1994.

Most preserved rollers are presented in pristine condition, but this does not reflect how they often looked in their working lives, particularly towards the end. Perhaps more typical of a tired working roller near the end of its days is this view of F class 10-ton compound roller, No. 10312, built in 1922. This was supplied to Portsmouth Corporation and is now preserved by Portsmouth Museum. It was having an outing to Netley Marsh, near Southampton, in July 2011.

Also museum-owned is No. 3319, dating from 1894. This is part of the collection of the Leicestershire Museum of Technology at the Abbey Pumping Station, Leicester. It is seen entering the premises after a run 'round the block' at an event in September 2012. Unlike the Portsmouth Museum roller it has no local connection with Leicester and was registered in Wiltshire.

A matching pair. NO 1224 (No. 9146) and NO 1221 (No. 8837) are BS class 10-ton rollers new to Essex County Council as one of ten. Similar NO 1218, NO 1225 and NO 1226 also survive. These two were seen together at the Enfield Pageant of Motoring in May 1997. On the front rolls is chalked 'Together again after 40 years'. NO 1221 worked for Essex CC until 1961, then sold off but was on contract to them for another two years before passing into preservation in 1964.

Many steam rollers were bought by contractors rather than local authorities. The largest of these was Eddison's Steam Rolling Co. of Dorchester with over 300 rollers at one time. One of their former Aveling rollers to be preserved in their identity is No. 10048 (FX 8307) of 1921, which worked until 1960. Rushmoor Arena, 30 July 1994.

Seen at the Chiltern Steam Rally, Prestwood, in 2004, No. 10431 of 1922 was one of the last engines purchased new by Eddisons. Most of its work was in the construction of airfields. Its last job was the construction of Heathrow Airport in 1944–48. It was converted into a traction engine in 1994 but was then restored back to roller form in 1997, receiving the Road Roller Association Trophy for preservation in 1998.

While most of the Aveling rollers preserved are in green livery, there are exceptions. No. 8097 (FX 7014) *Moby Dick* is another former Eddison engine, but this has carried a blue livery for most of its time in preservation. During the Second World War this was used to crush tin cans for recycling.

Some rollers owe their survival to having been placed in playgrounds after withdrawal. One such is No. 10671. Built in 1923 and displayed at the Newcastle Show, it was then sold to the Eddison Steam Rolling Co., Dorchester. In 1957 they presented this to Westminster City Council who located it in a playground on the Churchill Estate, Pimlico, London, until it was removed for restoration *c.* 1981. Appropriately it is now named *Churchill*. This view was taken at Knowl Hill in 1998.

There is a remarkable story behind this roller, No. 10114 (XD 8230). It was new to the Metropolitan Borough of St Pancras in 1921. In 1953 it was installed in a children's playground at Cumberland Market, London, NW1, as a climbing frame. After an 'incident' in the mid-1960s it was covered over in an earth mound as a base for a slide. It then became forgotten about until the London Borough of Camden decided to landscape the area in 1989 and it was rediscovered. After a feasibility study it was sent away for restoration in 1991. Before that it made this appearance on a low loader at Little Wymondley in 1990.

Restoration was completed by A. G. Bicknell & Co. at Hollycombe, and the roller returned to steam in 1995. Now named *Cumberland*, the roller is seen at Prestwood in 1999.

Steam rollers are not really designed for long-distance road running, but this has been no deterrent to the owners of DW 7088 *Rhoda*. In 2014, it had become the first steam roller to complete the journey from John O'Groats to Land's End. Here a year later in 2015 it completed the somewhat shorter London to Brighton HCVS road run. This was fitted with a new firebox in 2008 and also rubber tyres to the steel rolls to make a smoother (and quieter) journey. Built in 1930 as No. 14044, DW 7088 was new to H. Richards & Co. of Newport, Monmouthshire.

Celebrity Roller 1. Probably the best-known steam roller is DM 3079 *Betsy*, as this was owned by the former steeplejack and TV personality Fred Dibnah. No. 7632 was built in 1912 and was new to Flintshire County Council. Here he was with it a long way from his Bolton home at the 2001 Netley Marsh Steam & Craft Show near Southampton.

Celebrity Roller 2. No. 5590 (CH 3282) is the roller that had a starring role in the famous Ealing Studios comedy film *The Titfield Thunderbolt*, where it had a confrontation with the railway locomotive working the railway line the locals were trying to save. The display board in front shows some stills from the filming.

No. 12058 was built in 1928 and exported new to Australia. However, at the time there was an import tax on steam rollers in Australia. To avoid paying this, the engine was modified as a tractor when exported and then converted back to roller form upon delivery. Now repatriated, it has been given an age-appropriate British registration number BF 4501. Taken at Weeting in 2023.

Looking rather different to the average Aveling roller is No. 11423 *Smokey*. Registered KM 2229, this engine of 1926 was built as a tractor. It was later converted to a roller during its working life with a front end by John Allen & Sons, Oxford. It was seen at the Great Dorset Steam Fair (GDSF) in 1988. Since then, it has been rebuilt as a tractor, *c*. 2005.

Although Thomas Aveling perfected the three-point roller, not all the company's rollers were to this pattern. No. 6530 of 1908 is a 6-ton tandem roller of class AM6. When photographed here at the Expo Steam, Peterborough Showground, in 1980 it was in this form with no canopy fitted. It was then owned and had been restored by Tom Varley.

Two years on and NM 291 is again at Peterborough but in a new livery, and it has had a canopy fitted. There is also a matching living van.

NM 291 was the first steam roller bought by Luton Corporation. Being narrower than a conventional roller, these tandem rollers could roll between tram lines. Luton had introduced standard gauge trams in 1908, which lasted until 1930. NM 291 has now returned home, being housed at the Stockwood Museum in that town. Here it is at the Bedfordshire Steam & Country Fayre, Old Warden, in 1998.

A unique survivor is No. 7411 (XP 2915) *Chimaera*, built in 1911. This is the only remaining example from nine tandem rollers that were built with vertically mounted cylinders driving the rear roll through Shay gearing. This view shows the offside on which the encased cylinders and gearing are located. This was in a quest for a speedier change from forward to reverse when rolling asphalt.

The nearside of No. 7411 has a vertically mounted water tank. The roller has a tendency to wobble from side to side as it moves. These views were taken at Rushmoor Arena in 1988.

Another unique survivor is No. 12023 *Murphy* of 1928. This is a P type tandem roller, with a vertical boiler and twin horizontal cylinders. It also has power assisted steering on the rear roll. It spent its life with a quarry owner and road contractor at Portadown, County Armagh, Northern Ireland. Brought back to England as derelict in *c.* 1975, it was rebuilt in Hampshire with a new boiler between 1982 and 1984. The roller was seen at Havenstreet on the Isle of Wight in 1985.

Another view of the vertical boiler roller IB 2854. From 1987 until 2008 it was located overseas, in Holland and later in Germany, but it is now back in the UK and has been resident at the Amberley Museum in Sussex since 2013.

Two Aveling steam rollers in preservation are listed as having been converted to diesel power in later life. No. 10955 *Wexford Lady* was built in 1924 and converted later in its working life. This was seen at Carrog on the Llangollen Railway in September 2001. However, the *Traction Engine Register* records that this was returned to steam power in 2016. It also reports that this is now plinthed at the Wexford CC Depot, Ennisworthy, Ireland.

AX 3590 is an early roller, No. 2291 of 1887. However, this R10 model was converted to diesel with a McLaren engine in 1938. Now preserved in Essex, it was a visitor to Weeting in 2018.

Aveling-Barford (Aveling-Barford Ltd, Grantham, Lincolnshire)

Aveling-Barford moved production to Grantham and started a new numbering series from 1937. All of the rollers preserved in the UK are single-cylinder models. The first of these listed in the *Traction Engine Register* is No. AC601, CPT 233 *Talisman* of 1937. This W class 10-ton roller is seen at Prestwood, Buckinghamshire, in 1999.

No. AC624, 6-ton R class roller, was supplied to Coventry Corporation in 1938. It seems somewhat surprising that a roller built this late would not have a canopy fitted – the councillors probably didn't want to spend the extra money!

A similar roller, No. AC621, is suitably equipped. Now preserved in Essex, this was present at Barleylands in 2008. Note the small size of the flywheel on these Aveling-Barford rollers.

The last listed roller is No. AH425, an 8-ton T class engine built in 1948 and sold as late as 1950. JRM 966 *Sarah* was at Weeting in 2003.

Babcock & Wilcox (Babcock & Wilcox Ltd, Lincoln)

The five steam rollers listed in the *Traction Engine Register* for this company were all built by Clayton & Shuttleworth Ltd after that firm was absorbed by Babcock & Wilcox Ltd of Lincoln in 1924. Babcock & Wilcox applied their own maker's plates to the engines. All five are single cylinder engines. No. 95/4009 (YB 7976) is one of two 6-ton rollers. This was photographed at the Lincoln Rally in August 1997.

The other three surviving engines are of 10-ton weight. No 95/4012 was taken at Stoneleigh in 1987. This was new to W. W. Buncombe & Sons, Highbridge, Somerset, as I believe were the other survivors.

Burrell (Charles Burrell & Sons Ltd, Thetford, Norfolk)

Charles Burrell & Sons Ltd came from Thetford, Norfolk, so it is fitting to have a Norfolk-registered engine which is preserved in Norfolk. No. 3962 *Boxer* of 1923 is an 8-ton single cylinder roller that is part of the collection at the Bressingham Steam Museum near Diss. Note the full-length canopy.

Another Norfolk registered roller is No. 3047 *Sally*. This is an A class double crank compound engine weighing 10 tons. This was at the Great Dorset Steam Fair in 2000.

This 10-ton single cylinder roller No. 4041 (UO 793) was bought by the contractors R. Dingle & Sons of Stoke Climsland in Cornwall, whose lettering it retains in preservation. This was also taken at the GDSF in 2000.

A smaller 8-ton roller, No. 3973 *Fire Queen* dates from 1924. This comes from Essex and was taken at Marks Tey in 1989.

No. 3622 *General French* was built in 1914 as a Gold Medal 4 nhp tractor and used for removals. It was returned to Burrell's in 1927 and converted to an 8-ton roller, in which state it has remained. This was at Old Warden in 2004. It has been on loan to the Charles Burrell Museum at Thetford.

Like No. 3047, No. 3962 *Buster* worked for the contractor Doran Bros of Thetford. This is another resident at Bressingham but is unusual in being fitted with bi-directional scarifier. Seen at Weeting in 2018.

Clayton & Shuttleworth (Clayton & Shuttleworth Ltd, Lincoln)

Oldest of the steam rollers preserved in the UK is No. 34980 (XC 9578) *Excelsior* of 1902. This was new to A. Ward, Egham, and was sold for scrap in 1956. Taken at Carrog on the Llangollen Railway in 2001.

Above: No. 48835 (YA 7417) dates from 1923. This was taking part in the road run from the Birmingham Science Museum in 1984.

Right: No. 48946 (YA 9576) was built in 1924. After being exhibited at the Royal Show at Lincoln this 10-ton roller was supplied to W. W. Buncombe of Highbridge, Somerset. It was purchased for preservation in 1967 and is seen at Prestwood, Buckinghamshire in 1993.

Fowler (John Fowler & Co. (Leeds) Ltd, Leeds)

John Fowler is best known for having invented the system of steam ploughing with a pair of engines having winding drums mounted below their boilers pulling a plough back and forth between them. However, John Fowler & Co. (Leeds) Ltd made a vast range of engines including hundreds of rollers, their last being built in 1938.

Not all traction engines preserved have been restored and Fowler roller No. 17591 was displayed at the 2022 Great Dorset Steam Fair in 'as found' condition. A lot of time and money will be needed before this engine steams again.

Also unrestored at the time, Fowler No. 11095 of 1908 was displayed at Knowl Hill thus in both 1993 and here in 1994. Probably part of the collection of the late John Keeley the 2020 *Traction Engine Register* records the location of this as 'unknown'.

The immaculate condition of this engine, No. 16235 of 1924, is probably far from its normal working life state. A DN1 model, AV 45 *The Highlander* was new to Aberdeen County Council as a tar spraying roller. Now preserved in Essex, this was a regular attender at Barleylands where it was photographed in 2010.

No. 19481 (MAN 57) is a 12-ton DNC model dating from 1935. This originally worked on the Isle of Man where it was used for tar spraying. Here it is seen at Knowl Hill in 1983. Note the large belly water tanks. This roller has since changed hands and has been converted to a tractor with showman's fittings. It has also now been re-registered with a British registration as 497 UXE.

A smaller engine, No. 19032 *Doris* is a 1931 T3A model 6-ton roller. This was exported to Java where it spent its first eight years before returning to Britain. An unusual feature is gauges and plates giving metric measurements. Taken at Redhill in August 1994.

1929 DNA class roller No. 17560 spent its working life with Kitchener Bros, Potton, Bedfordshire. In 1961 this was bought by Reg Saunders of Stotfold and so began the Saunders family collection – now probably the biggest collection of operating traction engines and mechanical organs in the UK. Unlike the other engines in the collection this displays the Saunders family name and is seen here at Enfield in 1990.

A T3B class roller is No. 17951 (EGP 359) of 1931. This was photographed at Rushmoor Arena in July 1994. It was originally equipped for tar spraying.

Fowler No. 18659 *Arfur* is a DNB class roller of 1930. This spent its working life in Portugal, being returned in 1975. Taken here at Prestwood, Buckinghamshire in 1990, the roller has since been fitted with a canopy. It was named *Arfur* because it was bought by two people who owned 'arf' each.

EXX 760, No. 18843 *Princess Caroline*, is a class T3a roller from 1930. Later life saw it in use by the Mechanical Tar Spraying and Grouting Co., Reading. It was sold into preservation in 1959 and passed to John Crawley in 1961. This featured in the 1962 film *The Iron Maiden*, Old Warden, 2014.

No. 18874 (TK 5449) *Lord Jellico* was bought by Michael Oliver, founder of the Great Dorset Steam Fair, in 1968 but remained unrestored and derelict in his garden before being removed to attend the fiftieth anniversary of the National Traction Engine Trust at the GDSF in 2004. Restoration was started in 2005 but not completed before his death in 2009. Still owned by the family, it was seen at the Great Dorset Steam Fair in 2016.

The penultimate Fowler roller in the *Traction Engine Register* is No. 21833 (CRL 110), built in 1936. This is a 13-ton DNC model. This was seen at Ardingly in 1984.

A road-making display at the Woodcote rally in 2007 sees Fowler No. 17506 (UA 3582) *Undaunted*. A tar boiler is seen to the left. Unfortunately, this roller is no longer as such, having been sold to a new owner and converted to a tractor with showman's fittings in 2009 and is now named *Little Dawn*.

The rear view of UA 3582 shows the tar-spraying apparatus, retained from its working life with The Mechanical Tar Spraying & Grouting Co., Reading. Also participating in the display is No. 19049 (TK 6488).

A superbly recreated scene of period road making at the 2008 Woodcote rally. The roller is again No. 19049 (TK 6488) *Bacchus*, an 8-ton DNA class engine built in 1931. Its first owner was Frederick Sharp & Sons, Blandford, and was later sold to Bournemouth Corporation. Since first bought for preservation in 1968 it has passed through several owners and the owner at this time roaded it to all events it attended, covering over 500 miles in 2007.

TK 6488 *Bacchus* again. This time threading through the back streets of Birmingham with a cart and living van in tow. Only the modern lamppost and 'for sale' notice reveals that this is in preservation time and not back in working days. 13 May 1990.

A Fowler on the road – but in a very different context! No. 16148, TT 1253 *Ollie* was a participant in the 2010 London Lord Mayor's Show, representing the Worshipful Company of Paviors, one of the livery companies of the City of London who were responsible since medieval times for road maintenance. It was being followed by Marshall roller No. 80608, which was towing a float. The pair are passing the gates of the Royal College of Arms in Queen Victoria Street.

Above: No. 8111 (BW 4798) was built in 1898 by John Allen & Sons of Oxford, using Fowler parts. It is an example of the D2 class, weighing 12.5 tons. This was a visitor to the rally at Tallington, near Peterborough, in May 1997.

Left: The John Allen & Sons maker's plate on BW 4798. They were contractors and engine repairers who built several engines from Fowler parts but numbered in the Fowler series.

Garrett (Richard Garrett & Sons Ltd, Leiston, Suffolk)

Richard Garrett & Sons Ltd, Leiston, Suffolk, built most types of traction engines, and were particularly noted for their tractors, but rollers were a minor part of their production. Only eight are recorded as surviving in the *Traction Engine Register*. One of the oldest is No. 28408 of 1910. This is a 10-ton roller, which, as the canopy states, was in the fleet of the Cumberland County Council.

Seen at Barleylands, Essex in 2006, and visiting from its home in Cheshire, is No. 34084 (BJ 7045) *The Baroness*. This 10-ton roller dates from 1922 and worked for East Suffolk County Council.

Also at Barleylands, but in 1995 is No. 34267 of 1924. Note the different style of headstock on this roller.

Seen at Chatham Dockyard in 2004 is No. 34594. As with each of the others this is a 10-ton model. Registered in Ireland as IT 644, it is now resident back in the Irish Republic.

Green (T. Green & Sons Ltd, Leeds)

T. Green & Sons Ltd of Leeds specialised in steam and internal combustion rollers. Around 200 were made for the UK between 1872 and 1931. There are six surviving rollers dating from 1894 to 1925. TA 2431, No. 1968, dates from 1917 and is a single-cylinder 8-ton machine built for the War Department and then sold to Devon County Council. Seen at Tallington in 1992.

Henschel (Henschel und Sohn, Kassel, Germany)

Two rollers from this company are present in the UK. No. 5063 *Helga* was built as late as 1953 and was the last steam roller to be built in Germany. It was imported in 2004 and was at Preston, Kent, on 27 June 2009.

Iroquois (Barber Asphalt Co., Iroquois Iron Works Co., Buffalo, New York State, USA)

Above: In 2016 a most unusual roller making its first rally appearance was this vertical boiler tandem roller built by the Iroquois Iron Works, Buffalo, New York State, in 1920. This was imported in 2006 and restored by Mark & Luke Wingfield. This was photographed at Fawley Hill, the home of the late Sir William McAlpine, on the occasion of the Vintage Festival held to mark his eightieth birthday.

Left: A close-up view of the Shay gearing on the Iroquois roller.

Kelly-Springfield (Kelly-Springfield Road Roller Co., Springfield, Ohio, USA)

One of three rollers by this maker now preserved in the UK, the builder's plate is missing but it is believed to date from *c.* 1916. This is a nearside view of the roller. Note the under-boiler water tank. Preston, Kent, in 2009.

The offside of the roller. Note that while the steering wheel and rodding is on the nearside as in British practice, the flywheel is located on the offside.

Mann (Mann's Patent Steam Cart & Wagon Co. Ltd, Hunslet, Leeds)

Only one roller survives from this company, but it is an unusual design with similarities to the wagons built by this maker. There are two rolls at the front and one at the rear, intended for maintaining paths and drives. This was built in 1917 and sold to Kent County Council, later passing to Wingham Engineering. First restored in the 1950s, the present owners purchased it in 1989 and after an extensive rebuild it returned to steam in 2009.

The rear view shows the rear roll, also the chain drive and braking mechanism. Unusually, all rolls are mounted on springs – most rollers have no springs. Originally there would have been a body to hold asphalt ready for use. Both views taken at the Gloucestershire Steam & Vintage Extravaganza in 2010.

Marshall (Marshall Sons & Co. Ltd, Gainsborough, Lincolnshire)

While Marshall's built a vast range of traction engines, with portable engines predominating, there are few rollers surviving from before the 1920s. No. 57168 (BE 8186) dates from 1911 and is the second oldest listed in the *Traction Engine Register*. This was taken at Crich in 1984.

A pair of Marshall rollers at Quorn in 1985. FU 1326 is one of a batch of Q class 10-ton rollers – Nos 76795–8 (FU 1325–8) built in 1923, all of which survive.

PU 804 is No. 76913 and also dates from 1923. Seen in 1984 at Ardingley. Note the distinctive 'Britannia' badge on the headstock of this and the previous two rollers.

RF 1834, No. 81095, seen at Rushmoor Arena in 1987 is an S type 10-ton roller dating from 1926. The rear wheel spokes are a one-piece steel casting. These were piston valve compounds with a single eccentric for each cylinder instead of the usual double-link motion.

NR 6120, No. 79087 *Jane*, is another S type dating from 1925. When new this was exhibited at a Peterborough trade show and then purchased by Market Harborough Rural District Council. It worked until 1960 and then passed into preservation. This was visiting the GDSF from Wolverhampton in 1990.

S type 8-ton roller No. 79669 *Joan* was built in 1925 and was new to Midhurst Rural District Council. It passed to West Sussex County Council in 1934 and then remained in use until 1983, becoming the last steam roller to work for a council anywhere in the country. Still owned by West Sussex County Council, it was then retired into the care of the Amberley Museum where it now resides. In 2011 it was used to haul the float of the Worshipful Company of Paviors at the London Lord Mayor's Show.

In 1963 a group of six unrestored traction engines and rollers were stored at Manor Road Garage, East Preston, near Littlehampton. They would remain untouched there for forty-five years until 2008 when then owner Henry Lillywhite (whose father had stored them) disposed of them. One of these was Marshall roller No. 85740 (PL 1564) dating from 1930 and new to Wimbledon Council. This was removed in May, towed by a Fowler road locomotive to Amberley Museum where it was displayed in June. It later featured at a rally held at the Weald & Downland Open Air Museum at Singleton, West Sussex, before restoration commenced.

No. 85060 (VX 2427) *Porky/Dobbin* is an example of the U or 'Universal' class roller. This has two high-pressure cylinders with piston valves and no flywheel. There is instant reverse for rolling asphalt. It is an 8-ton machine dating from 1928 and new to Ilford Council. Taken at the GDSF in 2001.

The U class did not replace the S class as DV 4005, No. 85106, is an S class from 1929. The S class was built from 1923 to 1944. Taken at Prestwood in 2003, this roller was new to Devon County Council. Its last job was working on the Exeter bypass in the 1960s. Still owned by Devon CC, it is maintained by the Chiltern Marshall Preservation Group and was on loan to the Chiltern Open Air Museum.

Dating from 1931 is this 9-ton roller No. 86104 *Major Henry*. This was taken at Netley Marsh near Southampton in 2001.

Dating from 1935 is Q class No. 87635 (APX 542). Tallington, 1992.

The most recent surviving roller is No. 89549 (FRX 129) *Oor Wullie*, a 12-ton S class, built as late as 1944 when British production ended. Production continued in India for the Indian market after this date. Rushmoor Arena, 1986.

No. 87125 (VG 5715) is the survivor of only twenty-one of this VB class tandem roller built. Designed for rolling tarmacadam and asphalt, it can be instantly reversed and features steam-powered steering. This was new to Norwich Corporation in 1933. During the war years it was requisitioned for airfield construction. Sold off in 1946, it passed through several dealers before being purchased for preservation in a derelict condition in 1971. It returned to steam in 1987 and was rallied until 1998. It changed hands in 2006 and after overhaul returned to rallying in 2013.

VG 5715 seen again at the Great Dorset Steam fair in 1990, by which time a canopy had been added.

Robey (Robey & Co. Ltd, Lincoln)

The oldest conventional Robey roller is No. 32387 dating from 1913. New to an owner in Pembroke, this compound 10-ton roller is now owned by the Robey Trust Museum at Tavistock, Devon. It visited the Great Dorset Steam Fair in 2019 when the theme was engines made at Lincoln.

Most of the surviving later-built rollers are tandem rollers such as No. 42129 *Bullet* of 1924. This 8.5-ton roller features a stayless circular 'thimble' firebox as also fitted to some Robey wagons. This also appeared at the GDSF in 2019.

A rear view of the same Robey tandem roller.

Another example of this type is No. 41602, also from 1924, although this is a 7-ton model. Note the chain drive and underslung water tank. Preston, Kent, 2011.

Left: No. 42520 (FE 7632) *Barkus* of 1925 is part of the collection at the Bressingham Steam Museum in Norfolk where it was seen in 2021. The front roll on these rollers is made up from four sections as opposed to two sections on most other rollers.

Below: Three tandem rollers were rebuilt as tri-tandem rollers for Wirksworth Quarries Ltd. The rear rolls are chain-driven together. They were used in the construction of the M1 motorway. Two of these survive. These are Nos 44083 (VL 2370) and 45655 (VL 2773), both of which date from 1930. The latter is now part of the Robey Trust and this is also seen at the GDSF in 2019.

Ruston, Proctor (Ruston, Proctor & Co. Ltd, Lincoln)

PB 9976 is Ruston, Proctor No. 38591, a 6-ton SR class roller of 1910, and believed to be the only survivor of this weight. It was new to civil engineers G. F. Franks of Guildford. By 1958 it was laid up in the ownership of a Mr Selby, who offered it to Derek Marder of Andover for the sum of £25 or it would be sold for scrap. The engine has remained with the Marder family in preservation ever since. This view was taken at Knowl Hill in 1990.

BE 9143, No. 50735, is an SCR class compound cylinder 11-ton roller dating from 1915. This worked initially for Louth Borough Council, then various highway contractors until the early 1960s. Seen at Stoneleigh in 1988.

Ruston & Hornsby (Ruston & Hornsby Ltd, Lincoln)

When Ruston, Proctor & Co. Ltd merged with Richard Hornsby & Sons Ltd in 1918 to become Ruston & Hornsby Ltd they continued the existing makers numbers sequence of Ruston, Proctor. CJ 4853 is No. 114059 and is a SCR class 13-ton roller from 1921. In 1984 it attended the gathering at the Birmingham Science Museum.

Another example of the SCR class built by Ruston & Hornsby in 1922 is No. 115023 (XM 6373). As the canopy states this is now kept at the Bredgar & Wormshill Light Railway near Sittingbourne, Kent. This and No. 115123 (XM 6374), also now preserved, were new to Henry Woodham & Son, Catford. XM 6373 was sold for preservation in 1960 but dry stored until 1978. The present owners purchased it in 1988. Taken at the Essex Country Show, Barleylands, near Billericay, in 2009.

No. 158623 (UR 5657) *Midnight Magic* is an SR class roller of 1930. This was one of six made for an export order to East Asia which was cancelled. It was sold instead to A. T. Oliver, Wanden End, Luton, rolling local roads into the 1960s. This has been with the Vickery family since 2000 and has had a new boiler barrel and firebox. Here it was attending at Barleylands in 2001.

Taking part in the rolling demonstration at Barleylands in 2006 is XU 5721, an SR class roller from 1924.

Also representing the SR class is 1928 built No. 149813 (VF 3186) *Success*, here at Barleylands in 1991. This was new to Caxtons, Coltishall, and then with Plumbly & Gaze, Norfolk, remaining in service until 1962.

No. 52694 (NR 5056) of 1919 was new to Blaby RDC, passing to Leicestershire CC in 1930. Withdrawn in 1960, it was bought by Unilever Ltd and rebuilt with a diesel engine taken from a Nuffield tractor and gearbox from a Morris truck. It then remained in commercial use until 1987 and passed into preservation in 1988. Renamed *Nuffrush*, the converted roller is seen at Barleylands in 2009.

At the Essex Country Show at Barleylands in 2009 this remarkable line-up of six Ruston rollers was staged with *Nuffrush* on the right.

At the Bedfordshire Steam & Country Fayre in 2023 NR 5056 was on static display without its diesel engine. The roller has been purchased by the Vickery family, owners of Ruston & Hornsby rollers Nos 114059 and 158623, with the intention of converting this back to steam power.

Ruthemeyer (Ruthemeyer, Soest, Germany)

Three steam rollers by this German maker have been imported to the UK for preservation. This is No. 299 of 1919, a single crank compound 12-ton engine. This was on the premises of Preston Services in 2008. Alongside is a typically American traction engine, a 1910 built Port Huron.

SLM

Visiting the Weeting Rally in 1997 was this unusual steam roller from Switzerland built by SLM – Schweizerische Lokomotiv-und Maschinenfabrik. This was entered by the Dampfwalzer-Club Schweiz. The cylinders are mounted below the boiler on the substantial chassis frame and drive through railway locomotive type valve gear.

The other side of the SLM roller.

Tasker (W. Tasker & Sons Ltd, Andover, Hampshire)

No. 1933 (OT 8201) was the last steam engine made by Tasker & Sons in 1933. It was supplied to Watson & Haig, Acre Iron Works, Andover, and worked until 1962. It was then sold back to Taskers for their museum. It passed into the care of the Hampshire County Council Museums in 1972. Seen here at their display at the Rushmoor Arena Rally in 1994, it remains in working order and is normally displayed at the Milestones Museum, Basingstoke.

Also kept at the Milestones Museum where it is seen is No. 1906 of 1923. This is an 8-ton B2 class roller.

Trotter (A. W. Trotter, Gloucester)

This unique 1-ton roller was assembled at Coleford in the Forest of Dean by engineer Arthur Trotter in 1933. After his death in 1977 it was displayed in the Gloucester Folk Museum, making occasional rally appearances including at the GDSF in 2004. A new boiler was made and installed in 2006 but not commissioned and the roller was stored at the museum until 2016 when it was loaned to Jez Kirkwood for restoration to working order. This was completed in 2017 and here the roller is attending the 2018 Gloucestershire Steam & Vintage Extravaganza.

The roller has now been fitted with this water carrier tender, registered and certified for road use. It has also attended a wider range of rallies recently and this was it at the Amberley Museum Steam Rally in July 2023.

Wallis & Steevens (Wallis & Steevens Ltd, Basingstoke, Hampshire)

As the canopy states, HO 6139 *Big Emma* was built in 1896 for Andover Council and is the oldest surviving Wallis roller. No. 2357 is a 10-ton EXP class single cylinder roller. A well-known feature of rallies in Hertfordshire and Bedfordshire this was attending the Expo Steam at Peterborough in August 1985. It was originally fitted with Wallis & Steevens Patent Expansion Valve Gearing. This had been removed before the roller was rescued in 1968 but has since been remade along with a new boiler, firebox, and tender.

The second oldest is No. 2394 (BP 8585) *Caroline*, another 10-ton roller from 1900. Taken at Crich in 1984.

Slightly newer is No. 2574 (HO 6045) *Saracen* of 1902. This was bought new by Wrights of Alton and worked there for forty-seven years. It has been in preservation with the Rhodes family from Essex since 1976.

Seen attending a rally at Chartridge, Buckinghamshire, *c.* 1963 is No. 2656 (HO 5834) *Little Olga*. Built in 1903, this 5-ton, 3 nhp was built as a tractor and converted in 1920. This machine is no longer rallied in roller form, having been since rebuilt as a tractor. *Photo by Reg Batten*

A very similar machine is No. 2660 (CJ 4816), new to J. & G. Miller, Mitcham, Surrey. This was photographed at Havenstreet on the Isle of Wight in 1989. The *Traction Engine Register* reports that while this is based in the Southampton area, it is currently (2020) on long loan to the Blists Hill Collection, Ironbridge Gorge Museum.

While the previous five rollers were all single cylinder machines, No. 7742 *Clausentum* of 1921 is a two-cylinder compound. This was new to the City of Southampton and has remained in Hampshire after preservation. This was attending the 2011 Netley Marsh Steam & Craft Show, just a few miles from its old working area. Clausentum is the Roman name for Southampton.

No. 7745 of 1923 is a single cylinder 8-ton roller, which as the canopy indicates was resident on the Isle of Wight when photographed at Havenstreet in 1991. The registration ME 8584 is from Middlesex.

No. 7751 (HR 6354) dates from 1922 and is a class OB oilbath 8-ton roller. This was new to the Pitt-Rivers Estate at Rushmore, Wiltshire, where it was used for haulage and sawing as well as rolling. This was being rallied at the GDSF in 1997. This is believed to be the only W&S oilbath roller remaining as many others were converted to tractors.

A year on, 1924 saw the building of No. 7799 *Daisy May*, a compound 10-ton roller. This spent its working life in Hampshire, latterly with Southampton City Council. Note the cast wheels rather than spoked wheels, a feature shared with the later 'Advanced' model roller alongside. Knowl Hill, 1991.

The 'Simplicity' design was introduced in 1925, intended for the East Asian market. The distinguishing feature of these 3-ton rollers is the inclined boiler and circular stayless firebox. They were not a commercial success and only fifteen were made. OT 5187 *Pepperpot*, No. 7936, dates from 1927.

Six of the Simplicity rollers are listed in the *Traction Engine Register* and the most recent is No. 8023 *Christopher* of 1932. This is resident at Steam in the Country, Hollycombe, where it was seen in 1986.

The same engine after a repaint, attending the GDSF in 2001.

The prototype of the 'Advance' roller came out in 1923. This was designed especially for rolling asphalt. These had two high-pressure cylinders, one for forward and one for reverse with a toggle to switch between these giving almost instant reverse. The rear rolls were adjustable for camber. The standard canopy supplied was made of corrugated iron. Oldest of the surviving production models is No. 7784 (XP 2338). This is preserved in the Loughborough area and is seen appropriately at Quorn on the Great Central Railway.

The Advance design proved very successful. After 1930 all of the company's output was of this model until production ceased in 1940. This is No. 8005 of 1929, which we also saw during its working life (see p. 8). The steering is not by chains but is worked by a vertical wheel on the driver's platform via a geared quadrant mounted over the front roll. Taken in 2015 at the Gloucestershire Steam & Vintage Extravaganza.

PP 8588, No. 7922, dates from 1927 and spent its working life with Buckinghamshire County Council. It remained there in preservation with David Davis of Great Missenden. It was photographed just down the road from there at the Chiltern Steam Rally, Prestwood.

OT 7256 did not stray far from its birthplace at Basingstoke, being sold to the Borough of Eastleigh in Hampshire. This was a visitor to Barleylands in 1995. Note the canvas sheets which can be lowered from the canopy when the engine was parked up overnight.

No. 8105 was built in 1936. This is now a resident at the Amberley Museum where it is seen in 2023. Unlike most of the other rollers in this book this was not purchased by a local council or road-making contractor but saw military service with the Royal Engineers whose lettering it now carries.

The last steam roller made by Wallis & Steevens for the UK civilian market is No. 8112 of 1939. Now based in Kent, this is seen in East Sussex at the Wiston Steam Rally in 2015. Although registered in Hampshire, presumably by the makers or dealers the canopy indicates that this was purchased by Pontypridd UDC. It ran for them until 1962 when sold for preservation for £30, although the purchaser had to track down the last driver to buy back the missing brass items for another £5!

Zettelmeyer (Zettelmeyer, Konz bei Trier, Germany)

This company made a total of 632 steam rollers between 1910 and 1950 comprising four different models in weights from 8 to 16 tons. Diesel rollers were built from 1931 to 1978 and diesel vibration rollers from 1955 to 1982. The company was later acquired by Volvo.

Four rollers of this make are listed as preserved in Britain, plus one converted to diesel in Ireland. No. 553 of 1930 was imported from France in 1969. This is a compound 10-ton model.

No. 302 was built in 1926. In 2014 this was imported from Germany in pieces and rebuilt over the next four years. This was it at St Albans in 2017 when restoration work was still ongoing. Since this view, the boiler plates have been fitted and a canopy installed.

A year later and the boiler work is complete. It has been allocated British 'period' registration number BF 8138.

3
Convertible Rollers

As the most prolific of the steam roller builders, it is no surprise that Aveling & Porter also produced several convertible engines. No. 8679 (BT 8812) *Nellie Dene* is a convertible 6 nhp traction engine of class KLD. This is preserved in the Lincoln area, which is where it was seen in 1997.

Convertible engines were ideal for large estates and Aveling & Porter No. 5800 (BP 6043) was supplied in 1905 to the estate of J. Buchanan Esq., Lavington Park, Petworth, Sussex. He was the founder of the distillers James Buchanan & Co. Ltd as well as a racehorse owner and breeder. After his death in 1935 the estate was sold and the engine sold on. It was purchased for preservation in 1963, minus the rolls so is rallied in traction engine form. This was it at Netley Marsh in 1993.

No. 9181 (EP 1649) is a convertible 4 nhp tractor. Note the flange plate on the top of the smokebox on each of these engines for attaching the front roll.

No. 5541 *Bo-Peep* was built in 1904 as a tractor. It was then used as a roller from 1923 onwards and passed into preservation as such. It was converted back into a tractor in 2003 after this appearance at the GDSF.

Another engine that has been rallied in both roller and tractor form is Aveling & Porter No. 6319 *Queen of Herts*. This was built as a tractor in 1907 and sold to the Herne Bay Motor Company. It was fitted with rear rolls in 1928. When photographed at Knowl Hill in August 1983 it was running in roller form and painted maroon.

Moving on to Prestwood in 2004 and now No. 6319 is being rallied as a tractor and in green livery, being converted in 1991–92. The canopy lettering displays its credentials. In 2023 it was being rallied in a maroon livery again and without the smokebox flange ring.

At Old Warden in 2009 there was a theme of crane engines. With these on hand, the opportunity was taken to demonstrate how the switch could be made using crane engines (although in practice, jacks or sheerlegs would be more likely used). Aveling & Porter BP 7501 *Queen of Herts* has been propped up while the roller plate is attached, held in place by Fowler RF 6092 *Wolverhampton Wanderer.*

In the next stage, the front roll has been brought in by Burrell AB 8904 *Old Tim* and is being positioned in place before the tractor is lowered down by *Wolverhampton Wanderer* for the holding pin to engage.

It was not only Aveling & Porter who made convertible engines. YA 2095 *Dorothy* is Clayton & Shuttleworth No. 46817 built in 1914 as a convertible traction engine. It was used as a roller through most of its working life but is in traction engine guise in preservation. This is part of the Shuttleworth Collection at Old Warden where it was photographed in 2014.

Tasker No. 1409 (AA 2299) of 1909 is a compound B2 class engine. Built as a convertible tractor, conversion to roller form was carried out by Tasker's in 1930 and as an experiment it was fitted with a five-section front roll. This forms part of the Claude Jessett Collection at Tinkers Park, Hadlow Down near Uckfield where it is seen. The 'Little Giant' nameplate is a brand name fitted to Tasker tractors.

BP 9469 *The Mistress* is an 1890 convertible traction engine built by Wallis & Steevens. It was new to Charter Bros, Wisborough Green, and spent all it's working life in Surrey and Sussex mainly on threshing and sawbench work. It was rescued in a derelict state in 1983 and undertook a twelve-year restoration.

Wallis & Steevens No. 7650 was built in 1922 as a convertible roller and registered DL 2864 on the Isle of Wight. This is where it was seen in 1981 in tractor form but with the extended perch bracket for the front roller fork still fitted. This engine has since moved to Ireland.

4

Rollers Converted or Rebuilt in Preservation

Left: The highest number surviving Allchin engine, No. 3858 (NV 100) *Knapp* was first registered in 1931, the year the company closed down. It was built as a roller but rebuilt as a traction engine *c.* 1977. This appeared at a rally devoted to the marque held at Northampton in 1996 but has since been exported to France in 2012.

Below: A chunky and sturdy-looking machine is Aveling & Porter No. 8601 *Albion* of 1915. This BTD class compound 10-ton engine was reconstructed from a roller in 1994. It has also changed registration being originally registered TA 5317. This was seen at Harewood House, Yorkshire, in 1997.

Aveling & Porter No. 11728 *Tolerance* was reconstructed from a B class 6-ton roller by 1977. The attachment for the front roll perch bracket is clearly visible. Great Dorset Steam Fair, 1995.

Many of the conversions from rollers have also been given showman's fittings, as seen here on No. 10590 *Susan* of 1923. This conversion was carried out in 1997. Steam rollers and other road-travelling engines were normally fitted with solid flywheels, apparently seen as less of a distraction to horses, whereas agricultural traction engines had spoked flywheels.

Looking most unusual with these tyres was VN 2092 *Hunslet*. Aveling & Porter No. 14068 is described in the *Traction Engine Register* as a 4 nhp tractor rebuilt from a roller in 1970 and would later be rebuilt again in 2001 when it received conventional traction engine wheels. This was photographed at Gerrards Cross, Buckinghamshire, in 1972. At the time it was lettered as belonging to Robert Oliver Ltd, Bishop Auckland, Durham, so what was it doing in Gerrards Cross? It was listed as located at Chester-le-Street in 2012 but had relocated to Scotland by 2016.

Fowler No. 15970 was built in 1924 as a T3 class tar spraying roller for the Mechanical Tar Spraying & Grouting Co., Reading. Withdrawn in the mid-1960s, it then had several owners until a new owner in 1980 converted it to a tractor. Another new owner in 1990 converted it to a showman's tractor in 1992 with a dynamo found in a hedgerow that had come off a Garrett tractor. It is now named *D. C. Ver* as a tongue-in-cheek reference to it not being what it originally looked like. GDSF 1996.

Another substantial-looking machine with a massive cylinder block, Fowler No. 17287 *Oriana* was built as a DNB class 10-ton tar-spraying roller. This was reconstructed in 1981. Seen at Weeting in 1993, it has since also gained a canopy.

Another engine fitted with what appears to be tractor tyres is Fowler No. 16006 (SA 6370) *The Puddock*. This was rebuilt from a roller in 1986 and showman's fittings were added in 1992. Hailing from Aberdeen, this had ventured all the way south to Chester-le-Street for a rally where it was seen in 1993.

We saw Wallis & Steevens No. 2656 *Little Olga* earlier in the roller guise in which it was being rallied in the 1960s (see p. 73). This is it in the more familiar tractor form to which it was originally built and has now been rebuilt, here noted at Rushmoor Arena in 1995.

5
Motor Rollers

Some of the early internal combustion-engine rollers looked very much like their steam-driven counterparts, even down to a chimney for the exhaust. This Aveling & Porter version has the spoked wheels and chain steering similar to that of their steam rollers of the period. But this also has two spoked flywheels. Taken at the Prestwood rally in 1999.

Aveling & Porter No. 13033 (TL 3320) dates from the 1930s. The chain steering has gone, disc wheels have replaced spoked wheels, but there are still two flywheels and a copper-capped chimney. This was resident at the East Anglian Railway Museum, Chappel & Wakes Colne, when seen in 1989.

A later model Aveling & Porter, SB 5713 was new to Argyll County Council. Seen at Preston, Kent, in 2008 you can see that the chimney has now been dispensed with. The flywheel remains and is being used to drive a stone crusher. The other roller behind is a Wallis & Steevens.

This little roller is NK 1099, a 1914 Barford & Perkins type E. Barford would merge with Aveling & Porter in 1932 to become Aveling-Barford. Taken at Little Wymondley, 1990.

FOT 397, a 1947 Advance roller by Wallis & Steevens. Note the rather basic-looking exhaust pipe. Knowl Hill, 1991.

Contrasting rollers from Aveling & Porter and Wallis & Steevens displayed at the Great Dorset Steam Fair in 2019.

Acknowledgements, Bibliography and Further Reading

Thanks to Online Transport Archive for permission to use photos from the John Meredith Collection.

BATTEN, Malcolm — *Traction engine rallies: An appreciation over seventy years 1950-2019* (Barnsley, Pen & Sword, 2023)

BEAUMONT, Anthony — Steam at work (Newton Abbot, David & Charles, 1981)

COLBECK, Simon (ed.) — *Old Glory Archive vol 5: the colour files* (Yalding, Kelsey Publishing, 2021)

JOHNSON, Brian — *The Traction Engine Register, 13th edition* (Horsham, Southern Counties Historic Vehicles Preservation Trust, 2020 and earlier editions)

JOHNSON, Brian — *Steam traction engines, wagons and rollers in colour* (Poole, Blandford Press, 1976)

LOCKETT, David & ARLETT, Mike — *Traction engines: a colour portfolio* (Hersham, Ian Allan, 2002)

PRESTON, J.M — *Aveling & Porter Ltd* (Rochester, North Kent Books, 1987)

RAYNER, Derek — *Road rollers* (Princes Risborough, Shire, 1992)

SAWFORD, Eric — *Steam Power, Farm & Highway* (Burton-upon-Trent, Trent Valley, 1988)

Old Glory (Yalding, Kelsey Media) Monthly magazine published since 1988

Steaming (Redditch, NTET) Quarterly magazine of the National Traction Engine Trust

Vintage Spirit (Cranleigh, Steam Heritage Publishing) Monthly magazine published since August 2002

Rally programmes from events I have visited.